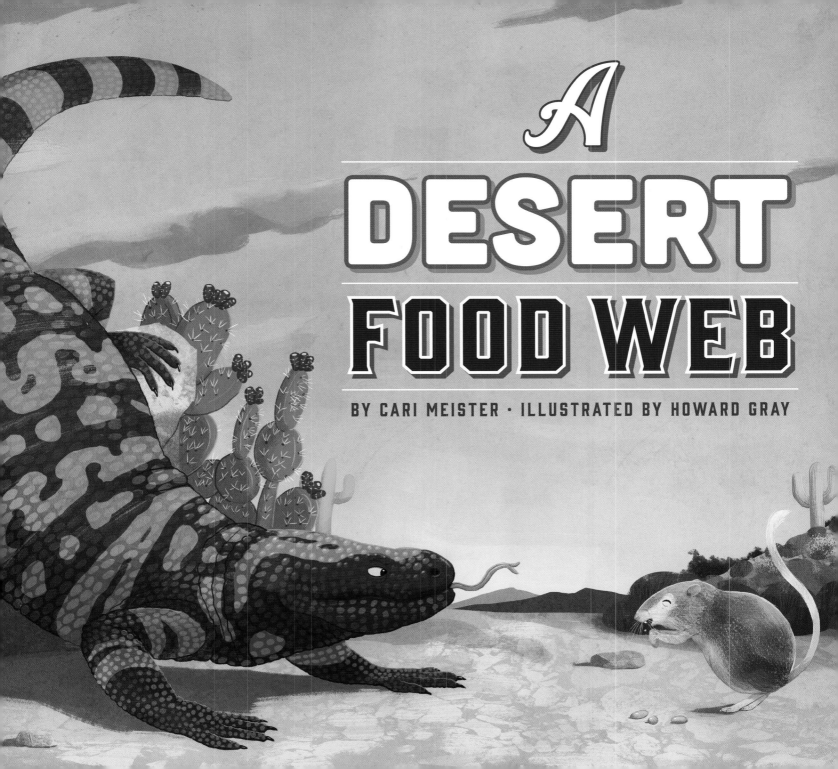

AMICUS ILLUSTRATED and **AMICUS INK**
are published by Amicus
P.O. Box 1329, Mankato, MN 56002
www.amicuspublishing.us

COPYRIGHT © 2021 Amicus. International copyright reserved in all countries. No part of this book may be reproduced in any form without written permission from the publisher.

Editor: Alissa Thielges
Designer: Kathleen Petelinsek

Library of Congress Cataloging-in-Publication Data
Names: Meister, Cari, author. | Gray, Howard (Howard Willem Ian), illustrator.
Title: A desert food web / by Cari Meister ; illustrated by Howard Gray.
Description: Mankato : Amicus, [2021] | Series: Ecosystem food webs | Includes bibliographical references. | Audience: Ages 6-9 | Audience: Grades 2-3 | Summary: "An illustrated narrative nonfiction journey to the southwest United States to show elementary readers how animals and plants in a desert ecosystem survive in an interconnected food web"— Provided by publisher.
Identifiers: LCCN 2019037206 (print) | LCCN 2019037207 (ebook) | ISBN 9781645490005 (library binding) | ISBN 9781681526423 (paperback) | ISBN 9781645490807 (pdf)
Subjects: LCSH: Desert ecology—United States—Juvenile literature. | Food chains (Ecology)—United States—Juvenile literature.
Classification: LCC QH104 .M454 2021 (print) | LCC QH104 (ebook) | DDC 577.540973—dc23
LC record available at https://lccn.loc.gov/2019037206
LC ebook record available at https://lccn.loc.gov/2019037207

Printed in the United States of America.

HC 10 9 8 7 6 5 4 3 2 1
PB 10 9 8 7 6 5 4 3 2 1

About the Author
Cari Meister has written more than 200 books for children, including the TINY series (Viking), and the FAIRY HILL series (Scholastic). She lives in the colorful mountains of Colorado with her husband and four sons. They enjoy desert mountain biking in nearby Moab, Utah. Cari loves to visit schools and libraries. Find out more at carimeister.com.

About the Illustrator
Howard Gray has illustrated a selection of fiction and non-fiction children's books. He has always considered himself an artist, but with a PhD in dolphin genetics, he has a background in zoology. He is now pursuing his dream career in children's illustration from the picturesque city of Durham, UK. Find out more at www.howardgrayillustrations.com.

A round-tailed ground squirrel runs along the rocky desert soil. It is spring. But it is already very hot in the Sonoran Desert in Arizona.

3

Temperatures during the day can reach above 100°F (38°C). It is very dry and rarely rains, but life here has learned how to survive.

The squirrel runs up a barrel cactus. It drinks the nectar to get energy.

All living things need energy, which comes in the form of food. When one living thing eats another, energy is passed along in a food web.

The sun shines down on desert plants, like the prickly pear cactus. Plants are producers. They can make their own food using the sun's energy.

All animals—from big to small—are consumers. Herbivores eat plants for energy. They are primary consumers. The squirrel dashes across the desert searching for fruit and grass to eat.

Is that a snake? The ground squirrel quickly runs off. It is lucky this time. A rattlesnake is venomous—one bite can be deadly.

A rattlesnake is a secondary consumer. It eats other animals for energy, but it also might get eaten. The squirrel is not its only source of energy. The snake also eats rats and mice. Its next meal doesn't get away. Gulp!

Dusk comes to the desert. It is cooler. Animals that have spent the hot day sleeping are now waking.

Whoosh! Here comes a great horned owl! It spotted the snake from above. It grabs the snake with its talons.

Look! A Gila monster comes out of a burrow. It flicks its tongue to smell the air. As a carnivore, it eats other animals for food. Nearby, a kangaroo rat nibbles on some seeds. Before the lizard can attack, a gray fox darts out. It catches the rat. That's dinner for the fox!

Don't worry, the Gila monster will find another meal. There are a lot of prey animals. Rats outnumber the lizards and foxes. But these rodents need to eat, too. If the grass disappeared, so would the rats. Then the fox and lizard would go hungry.

Which desert animals are at the top of this food web? The cougar and the golden eagle! They are apex predators. So is the great horned owl. Other animals usually don't hunt them.

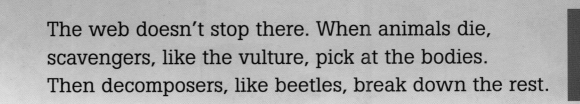

The web doesn't stop there. When animals die, scavengers, like the vulture, pick at the bodies. Then decomposers, like beetles, break down the rest.

19

Nutrients from dead animals and plants go back into the ground. This helps new plants grow and feed more animals. What an amazing desert web of life!

A DESERT FOOD WEB

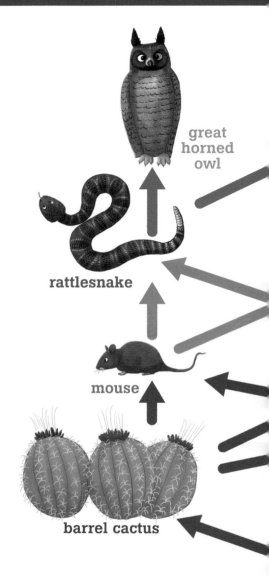

apex predator An animal that has no natural predators; it is the top consumer in a food web.

secondary consumer An animal that can eat both plants and other animals, but also gets eaten by other animals.

primary consumer An animal that eats only plants and is eaten by other animals.

producer A plant that makes its own food.

scavengers and **decomposers** Animals that break down dead animals and plants by eating them, which help nutrients go back into the ground.

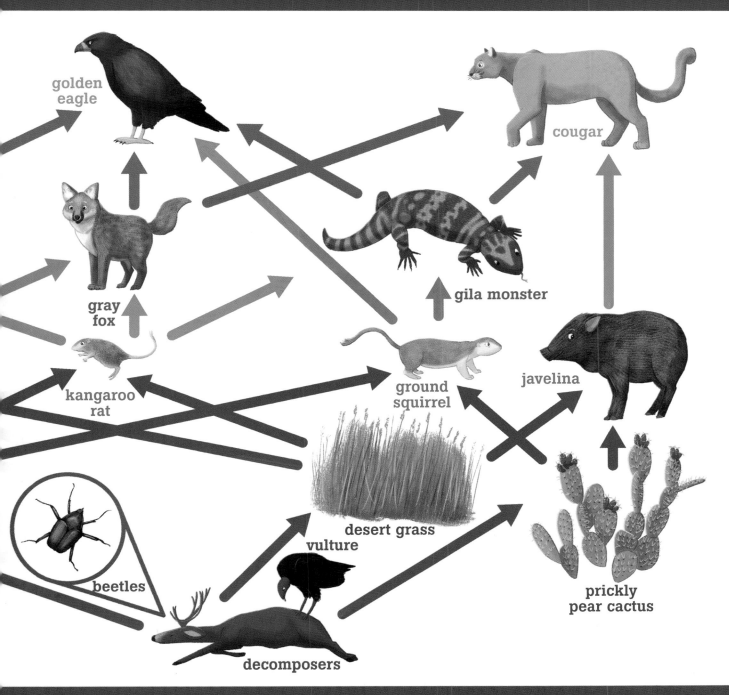

GLOSSARY

carnivore An animal that eats other animals for energy.

dusk The time of day after the sun goes down but it isn't completely dark.

food web An interconnected system of animals and plants within an ecosystem that shows who eats what.

herbivore An animal that eats plants for energy.

nectar A sugary fluid that plants make.

nutrient A substance that plants and animals need to live and grow.

WEBSITES

ASDM Sonoran Desert Digital Library for Kids
http://www.desertmuseumdigitallibrary.org/kids/Games.html

Desert Exploration | Anza-Borrego: Just for Kids
http://www.abdnha.org/just-for-kids/anza-borrego-just-for-kids-animals.htm

Every effort has been made to ensure that these websites are appropriate for children. However, because of the nature of the Internet, it is impossible to guarantee that these sites will remain active indefinitely or that their contents will not be altered.

READ MORE

Jacobson, Bray. **Food Chains and Webs**. New York: Gareth Stevens Publishing, 2020.

Pettiford, Rebecca. **Desert Food Chains.** Who Eats What? Minneapolis: Jump!, 2016.